MAPS,
THE OCEANS & CONTINENTS:
THIRD GRADE GEOGRAPHY SERIES

SPEEDY
PUBLISHING

Speedy Publishing LLC
40 E. Main St. #1156
Newark, DE 19711
www.speedypublishing.com

OCEAN

Around 71% of the Earth's surface is covered by oceans. Though they are all interconnected, they are generally divided into 5 Oceans, namely, Pacific Ocean, Atlantic Ocean, Indian Ocean, Arctic Ocean and Antarctic Ocean.

The Pacific Ocean is the largest on earth, it covers around 46% of the Earth's water surface. The Pacific Ocean contains around 25000 different islands.

The Atlantic Ocean is the second largest ocean on earth, it covers over 29% of the Earth's water surface. The Bermuda Triangle is located in the Atlantic Ocean.

The Indian Ocean is the third largest ocean on earth, it covers around 20% of the Earth's water surface. The average depth of the Indian Ocean is measured at 3,890 meters.

The Antarctic Ocean comprises the southernmost waters on the planet. Large icebergs are very common in its waters.

The Arctic Ocean is the smallest and shallowest of the world's five major oceanic divisions. It is located in the Northern Hemisphere and mostly in the Arctic north polar region.

CONTINENT

The continents are the planet's large, continuous landmasses. Conventionally earth has been divided into 7 continents.The list includes Asia, Africa, Europe, North America, South America, Australia and Antarctica.

Asia is the Earth's largest and most populous continent, located primarily in the eastern and northern hemispheres.

Europe is a continent that comprises the westernmost part of Eurasia. Europe is the world's second-smallest continent by surface area.

Africa is the world's second-largest and second-most-populous continent. Africa hosts a large diversity of ethnicities, cultures and languages.

North America is the third largest of the seven continents. North America covers a land mass area of about 9.5 million square miles.

South America is a continent located in the Western Hemisphere. South America is the fourth largest continent in size and the fifth largest in population.

Australia is the smallest continent by size and the second smallest in terms of population. Australia is surrounded by the Indian Ocean and the Pacific Ocean.

Antarctica is Earth's southernmost continent. Antarctica is the coldest, driest, and windiest continent.